A New Angle on Trains and Train Stations

BY SARAH MASTRIANNI

Copyright © Gareth Stevens, Inc. All rights reserved.

Developed for Harcourt, Inc., by Gareth Stevens, Inc. This edition published by Harcourt, Inc., by agreement with Gareth Stevens, Inc. No part of this publication may be reproduced or transmitted in any form or by any means, electronic or mechanical, including photocopy, recording, or any information storage and retrieval system, without permission in writing from the copyright holder.

Requests for permission to make copies of any part of the work should be addressed to Permissions Department, Gareth Stevens, Inc., 330 West Olive Street, Suite 100, Milwaukee, Wisconsin 53212. Fax: 414-332-3567.

HARCOURT and the Harcourt Logo are trademarks of Harcourt, Inc., registered in the United States of America and/or other jurisdictions.

Printed in China

ISBN 13: 978-0-15-360249-8
ISBN 10: 0-15-360249-X

8 9 10 0940 16 15 14 13
4500409990

CHAPTER 1:

Angles, Angles Everywhere

Before there were cars, buses, and airplanes, people traveled long distances by train. Powered by steam, diesel fuel, or electricity, trains have carried passengers and freight from one place to another for more than 100 years.

Many people have seen these connected cars that run along rails on tracks. The train station itself, however, may not be a common sight to most people. Train stations are also called railway stations. They are large, small, and nearly every other size imaginable. No matter the size, railway stations are usually bustling with people and activity.

A train station is one place to see geometric figures.

Train stations are the key to train travel. They provide a stop for both passengers and freight. Each station has information about the trains that run through the station and the times the trains stop. This information is called a schedule, and it is available on paper, on the Internet, and on monitors and boards at the stations.

Train tracks run north, south, east, west and all directions in between. Trains make scheduled stops at many stations along the way. The tracks run between towns, cities, and states, and across the country. Most trains don't just carry freight and passengers, though—they also carry geometric figures!

This is Union Station in Kansas City, Missouri.

Geometry is a branch of mathematics that deals with geometric figures. While math class is certainly one place to study geometry, a train station is another. Trains and train stations are loaded with lines and angles. They are a great place to find plane figures, too. A plane figure is a figure that lies in one plane. Rectangles and triangles are examples of plane figures.

Nearly everything you see in and around a train station contains geometric figures. Many stations have magnificent architecture with intricate details and design. Take a look at the photo on this page for example. The roof and windows create geometric figures.

Think about angles as just one type of geometric figure in a train station. Angles are made up of a point and two rays that extend from the point. A ray may look like a line, but lines continue forever from both ends. A ray is a part of a line that begins at one endpoint and extends forever in only one direction.

The large clock at Grand Central Terminal in New York City forms new angles with the tick of each minute. At 3:00 for example, the clock hands make an L. At 1:00 the hands form a lesser angle; at 5:00 the hands form a greater angle. The angle created at 3:00 on the clock is a right angle. The angle that appears at 1:00 is an acute angle and the angle formed at 5:00 is an obtuse angle.

The hands of the clock at Grand Central Terminal in New York City form different angles depending on the time shown.

This is the inside of Grand Central Terminal in New York City. There are many angles to see inside the building.

The clock at Grand Central Terminal is not the only place to see angles. The building itself has many angles. Notice that the rectangular window frames in this photo form right angles at each corner. Even the smaller glass panes within the larger frames form right angles.

Look at the counters and schedule boards surrounding the ticket windows. They form angles, too. If you look closely, you may see angles on the information booth at the center of the image. Look closer still, this time at the people and the shadows they cast upon the floor.

Trains on a track are also a great place to find angles. The windows on the train in the photo on this page form angles. Even the painted design on the engine and cars forms angles. Can you find any right angles in the photo? Can you find any acute or obtuse angles? Be sure to look at the train and the tracks.

What other types of geometric figures can you find on a train or in a station? One answer to that question is lines—all kinds of lines! In photographs and artwork, there are many designs that look like lines. They are straight marks, but they end at some point. Look for designs that look like lines in the photo.

Look for acute, obtuse, and right angles on the train and the tracks.

CHAPTER 2:

LINES — THEY JUST KEEP GOING

Some lines are parallel. Parallel lines are lines in the same plane that stay exactly the same distance apart. Some train tracks run parallel to one another. Engines pull cars along the track by riding the rails. Trains can run on parallel tracks at the same time without crossing paths.

The two metal rails that run along the inside of a track are always parallel. The rails must stay the exact same distance apart even if the track curves around a corner or over the surface of the earth. The width between the rails is called the gauge. The United States standard railroad gauge is 4 feet, 8.5 inches. More than half the tracks around the world are standard gauge.

Lines that cross are said to intersect. Two lines that cross are called intersecting lines. The point at which these lines meet is the point of intersection. Train maps show many points of intersection as tracks intersect one another. Just like lines in math, some train tracks are parallel and others intersect.

Here, the Pink and Green Routes are parallel. The paths of the trains never intersect. The Pink Route and the Red Route do intersect. Their tracks meet at North Way, their point of intersection. Wilson Boulevard is the point of intersection for three routes. The Green, Blue, and Red Routes meet at Wilson.

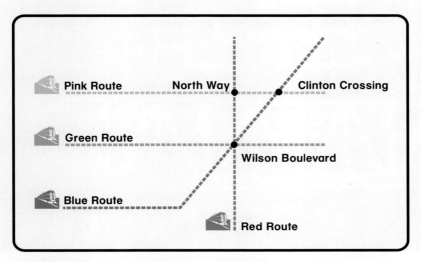

Parallel, perpendicular, and intersecting lines are shown on this train map.

Perpendicular lines are special types of intersecting lines. Four right angles form where perpendicular lines intersect. Railroad Crossing signs show perpendicular marks. You probably have seen this sign where a train track crosses a road. Tracks running north and south intersecting with tracks running east and west are perpendicular, too.

Perpendicular designs are everywhere in the photo on this page. Don't look at just the train. Can you find perpendicular designs on the bridge? The buildings have designs that remind us of perpendicular lines, too.

Perpendicular lines can be found on the building and on the train bridge.

This is Union Station in St. Louis, Missouri.

Angles and designs that remind us of parallel and perpendicular lines are nearly everywhere around trains and their stations. "All aboard," because there are still more types of geometric figures to find on trains and in stations. Plane figures are nearly as plentiful as lines!

The photo on this page is of Union Station in St. Louis, Missouri. Can you see plane figures in the image? Look at the railing along the bridge. Now look at the steel framework near the top of the station. Name the plane figures you see.

CHAPTER 3:

Plane Figures: Anything But Plain

Tiles on the train floor form shapes. Windows form figures, too. Even the dials that a train engineer uses form figures. Train conductors spend much of their day looking at shapes. The tickets they collect are shaped like rectangles, and conductors often punch a hole, sometimes shaped like a circle, into the tickets. The circle tells the conductor when and where a passenger will leave the train.

Riders on a train enjoy the many plane figures surrounding their seats and on the inside of the train. They also get to watch plane figures in scenery from the window as the train travels from stop to stop.

Quadrilaterals can be found on this train bridge in New York.

Steel train bridges are ideal places to spot plane figures or just plain figures! Steel bridges are very strong and sturdy. Over the years, steel bridges have replaced older, wooden ones.

Quadrilaterals are one type of plane figure. You can find quadrilaterals on steel bridges. Two common quadrilaterals are squares and rectangles. A square is a rectangle with four sides the same length and four right angles. A rhombus is another type of plane figure. A rhombus is a parallelogram with four congruent sides and congruent opposite angles. Where do you see rhombuses in the photo on this page?

This is Union Station in Washington, DC.

Plane figures are everywhere at Union Station in Washington, DC. It might seem like many train stations around the country are named Union Station. A union station is a large railway station with many tracks and different railway companies. This key station is named Union Station in many cities.

There are many examples of plane figures and angles in this photo of Union Station in the nation's capital. Look carefully from ceiling, to floor and notice everything from the octagons in the ceiling arches to the angles in the floor tiles if you imagined rays running from the tiles' black "points."

Take a train the next time you travel. Spend time looking for geometric figures. What plane figures do you see in the station? Point out right, acute, and obtuse angles before you climb aboard your train. Where do you see parallel and perpendicular designs that remind you of lines?

Once on your way, check the train car for more plane figures, line-like designs, and angles. Take a moment to look out your window. Can you see figures in the scenery before it passes from view? Don't just read about geometry in your math book. Take a trip on a train, or if you can't ride the rails, then visit a station. You are sure to see geometry in action!

Caption

This mural is displayed at Union Station in Los Angeles, California.

Glossary

acute angle an angle that measures less than a right angle

architecture building designs and making structures

intersecting lines two or more lines that cross at exactly one point

obtuse angle an angle whose measure is greater than the measure of a right angle but less than a straight angle

point of intersection the exact point at which lines cross each other

quadrilateral a polygon with four angles and four sides

rhombus a parallelogram with four equal, or congruent, sides

right angle an angle that forms a square corner

Photo credits: cover, pp. 5, 6, 7 courtesy of Metro-North Railroad; pp. 1, 2, 8, 12 © Hemera Technologies Inc.; p. 3 Mario Tama/Getty Images; p. 4 © Richard Cummins/Corbis; p. 10 © Kelly-Mooney Photography/Corbis; p. 11 © Lee Snider/Photo Images/ Corbis; p. 13 © David Zimmerman/Corbis; p. 14 © Catherine Karnow/ Corbis; p. 15 © Ted Streshinsky/Corbis.